GRAPES *AND* GRAPE VINES *OF* CALIFORNIA

Harcourt Brace Jovanovich, Publishers

New York and London

Library of Congress Cataloging in Publication Data

Main entry under title:
Grapes and grape vines of California.
Originally published under the auspices of the California State Vinicultural Association:
San Francisco: E. Bosqui.
Reprint. Facsimile originally published:
San Francisco: J. Windle Pub. Co., 1980.
1. Grapes—California—Varieties.
2. Grapes—Varieties—Pictorial works.
3. Viticulture—California.
SB389.G77 1877a 634'.83'09794 81-4775
ISBN 0-15-136786-8 AACR2

Printed in the United States of America

B C D E

INTRODUCTION

by Kevin Starr

Mid-nineteenth-century San Francisco was a city noted for its fine printing. Flung down on the far shores of the Pacific, a continent away from the printing centers of the East, San Francisco was forced to do its own printing work. Mysteriously, the decision was made to do this work with a gracefulness born of simplicity. The work of such firms as Edward Bosqui and Company, Towne and Bacon, and Charles A. Murdock and Company — to name a few of the City's more significant firms — stood in dramatic contrast to so much of the appalling work of the mid-Victorian era. Then, printers gorged themselves on founts of execrable type faces, and cluttered vast reams of perfectly innocent paper with typographical confusion. Mid-Victorian printing shows the machine run amuck. Surrendering itself to industrial values, printing forgot its ancient heritage as a craft and a visual art.

The San Francisco style, by contrast, was simplicity itself. The printers of the City had a taste for pre-industrial type faces, Franklin Old Style being among their favorites. They used these graceful forms in conjunction with engravings, generous spacing and margins, and an uncluttered layout. As a flourishing art form, printing in San Francisco achieved a style which combined assertive modernity with a subtle nostalgia for printing as a pre-industrial art. Through Franklin Old Style, the printers of San Francisco recalled that time when printing existed as an art and a craft, not just as an industrial trade. And yet, the elegance so often evident in even the most routine commercial printing done by these San Francisco firms declared that even the machine could be made to serve beauty.

The dean of these mid-Victorian printers was Edward Bosqui (1832–1917), a Montreal-born Quebecoix who arrived in San Francisco in 1850 as a lad of eighteen, eager for the gold fields. Bosqui found his destiny not in the Mother Lode, however, but on the corner of Clay and Leidesdorff Streets — as proprietor and aesthetic architect of Edward Bosqui and Company, established 1863 and incorporated some years later, in 1881, as Bosqui Engraving and Printing Company. Bosqui's firm occupied the entire top floor of a building that ran from Clay to Commercial on the east side of Leidesdorff. These same quarters were destroyed by fire in 1893, rebuilt, then destroyed again in April 1906, a catastrophe from which Bosqui never professionally recovered. Nevertheless, within these twenty-five years, Bosqui issued some of the most beautiful books ever printed in the history of this country.

Edward Bosqui was one of those bluff, hearty bohemian businessmen — lovers of learning and enterprise — who brought such cumulative verve to the early decades of San Francisco's commercial history. Art and commerce coexisted most fruitfully in Bosqui's Gallic soul. On the one hand, he was, like the rest of his generation, determined to make a fortune in California. Bosqui's *Memoirs,* privately printed in 1904 in a limited edition of thirty copies, bristle with accounts of Bosqui's entrepreneurial enterprises in banking, mining, ranching, real estate speculation, whatever. He hardly mentions his printing career, which does not begin until some fourteen years after his arrival in California — when a number of other easier avenues to wealth had temporarily exhausted themselves. To the end of his life (he died on 8 December 1917 at his home at 2010 Lyon Street, San Francisco) Bosqui dabbled in a number of speculative ventures. He was, after all, a son of the 1850s, San Francisco's age of iron and ambition, when men wore leather aprons and kept stores that were the foundations of great fortunes.

Bosqui approached printing first and foremost as a business. He printed bank reports, legal briefs, bonds, certificates, church proceedings, mining and insurance company reports, wine and fruit labels, bank checks, wedding announcements, architectural drawings, restaurant menus, concert programs — as well as some 138 books and ambitious pamphlets. He did his best typographical work, moreover, at his busiest period of commercial success. Scholars of fine printing tell us that Bosqui's work declined artistically during the early 1870s when his business fell off because of recession.

Business, then, was in a very real sense the first source of energy underlying Bosqui's art. But more explicitly imaginative ambitions played their part as well. Edward Bosqui loved art, learning, and the life of the mind. Forced to leave school at an early age, he read and collected books in English, French, and Spanish, his three most fluent languages. In March of 1871 he helped to found the San Francisco Art Association; and the next year, in April 1872, he and a group of like-minded journalists, artists, and bookish businessmen founded San Francisco's famed Bohemian Club. As a director of the Bohemian Club, Bosqui allied himself with the more bohemian element of the Bohemians — even when these improvident, often impecunious gentlemen of the garret and composing room were sadly in arrears for their Club bar bills. Bosqui's summer retreat in Ross Valley, Marin County, was filled with the paintings of his Bohemian friends — Thomas Hill, William Keith, Virgil Williams, Julian Rix, Toby Rosenthal — alas, paintings all destroyed in 1897, along with many of Bosqui's best books, in a disastrous fire, the second of three fire catastrophes Bosqui was forced to endure in his lifetime.

It was this aesthetic side to his nature, then, that prompted Bosqui's motto, "Quality, not quantity," when it came to the printing of books. Bosqui imported the recently revived Old Style type faces into San Francisco. He secured English Caslon type (in eclipse from 1800 to 1859) from the Johnson Type Foundry in Philadelphia. He imported Franklin Old Style from Boston. Bosqui had an almost Asian intuition of the value of negative space as a factor in design. He never cluttered his pages. Type and creamy white paper complemented each other with exquisite harmony, interspersed, when appropriate, by skillful engravings or bold initial letters ablaze in color. Bosqui endowed his pages with that graceful simplicity which is the essence of great typography. A Bosqui page served the content of its text through a direct, uncluttered presentation that was a statement about space, spatial relationships, and design harmony.

How precious are these Bosqui books to those of us who today still revere the achievement of that grand epoch of San Francisco printing! Have any *Custom House Regulations* of any city ever been printed so beautifully as Bosqui printed San Francisco's in 1864? Did the work of any young poet ever receive such a lavish debut presentation as Bosqui afforded the *Poems* of Charles Warren Stoddard in 1867? Even today, Stoddard's *Poems*, published by A. Roman and Company, delight us by a perfect interaction of Caslon, Franklin Old Style, etchings by William Keith, assertive initialing, and negative space; by its wine-red cloth binding with beveled edges, stamped so sumptuously in gold. Looking for favorites, one wanders delightedly through the canon of Bosqui imprints: to Mrs. Louise (Palmer) Heaven's novel,

In Bonds (1867), for instance, or to the *Poems* (1868) of John R. Ridge, the mystical half-Cherokee writing under the name of Yellowbird. Praise has been constant for the *Noticias de la Nueva California* by Padre Francisco Palou. Bosqui produced these four lavish volumes in 1874 for the California Historical Society, with Mr. Joseph A. Donohue, a San Francisco banker, hotelman, and pioneer in public utilities, picking up the $2000 tab. In 1896 *The Inland Printer* described Palou's *Noticias de la Nueva California* as "the best single specimen of bookmaking ever produced in San Francisco."

All this brings us to Bosqui's triumphant folio *Grapes and Grape Vines of California,* published in 1877 under the auspices of the California State Vinicultural Association. From the moment this extravagant imperial folio appeared, it stunned one and all with its dazzling presentation of full-page lithographic plates, lifelike in color, with opposing texts of Franklin Old Style. At the Paris Exposition of 1878, *Grapes and Grape Vines of California* astonished the judges. They simply could not believe that something so perfect could be produced in far-off San Francisco. Back home in San Francisco, bibliophiles rushed to offer their endorsement. Irving M. Scott, President of the Art Association, and Virgil Williams, Bosqui's artist friend and fellow Bohemian, offered joint praise. They considered *Grapes and Grape Vines of California* "among the most successful efforts of oleographic art ever produced in Europe or America. The bloom, the delicate warmth, and transparent glow of the natural clusters are rendered with wonderful fidelity to nature, and vividly portray the natural bunches, as we see them growing in the vineyards of California." Two decades later, William Loy, writing in *The Inland Printer,* described *Grapes and Grape Vines of California* as "probably the best and most elaborate specimen of bookmaking ever attempted in San Francisco."

Oddly enough, this undisputed masterpiece appeared just as the overall quality of Bosqui's work was going into decline. A candle, however, often flares into renewed incandescence before it dims; history is replete with examples of masterpieces achieved in the penultimate moments of their genre's vitality. *Grapes and Grape Vines of California* is an enduring book because it was made to be beautiful — and because it was made to serve a great enterprise, that of wine in California. It is more than just one of the most lovely books ever to be produced in California. This exuberant folio is also a significant moment in California's nineteenth-century journey to self-awareness as a regional civilization. It is a symbolic presentation of California as an American Italy, a Mediterranean shore, bathed in sunlight, carpeted in vines.

Obviously symbolic of civilization, the vines planted by Spaniards suggested to those Americans who first saw them that California was not an unrelieved wilderness. Like the missions themselves, the vines of California bespoke history, solicitude, and patience. Frenchmen joined in the growing of them, and then Americans. In the early 1860s, sensing California to be on the verge of an exciting new industry, the legislature appointed Count Agoston Haraszthy as Commissioner on the Improvement and Growth of the Grape-vine. Under instructions from the legislature, Haraszthy traveled through Europe between 1861 and 1862, compiling information and sending back about 100,000 cuttings for experimentation. In *Grape Culture, Wines and Wine-Making* (1862) this enterprising Hungarian émigré provided California with the contours of a great dream: hills covered by vineyards, from which went forth to the world wines which were mellow and yet, like California, had a pleasing young strength. By the late 1860s part of the Count's dream was coming true. Vines covered the once empty hills. In the winter of 1867–68 over 200,000 were planted in Sonoma County alone. Napa County had over a million vines planted by the end of the decade. Tended by Frenchmen, Mexicans, and those Southern Europeans whose ways of living were easily transplanted to this congenial climate, the vineyards conferred upon coastal California a Mediterranean texture.

Throughout the 1860s and 1870s phylloxera (plant lice) ravaged the vineyards of France. Planting vines in the millions, Californians could thus feel themselves engaged in an act of preservation and renewal, offering new life in a new soil under another sun to the endangered vines of Europe. It was a startling, wonderful belief, one that put Californians in the service of history; and they bore witness to this new role in the most beautiful book ever produced in a city noted for its fine printing.

Some time in 1875 the California State Vinicultural Society commissioned Miss Hannah Millard to do a series of watercolors illustrating the varieties of grapes grown in California. Millard produced ten such paintings, brilliantly colored and perfect in detail. For promotional purposes, the Vinicultural Society asked Bosqui to reproduce Millard's paintings for wider distribution.

Bosqui had in 1876 only just added lithography and engraved printing to his repertoire. He secured the services of William Harring, formerly of L. Prang and Company, Boston, to transfer Millard's watercolors to lithographic stones. A committee of Bosqui's artist friends — Toby Rosenthal, William Keith, Jules Tavernier, William Bradford, and Julian Rix — sat in as an advisory committee to insure fidelity of detail and perfection of color. They and William Harring did their job superbly, as all found out that exciting day when the first sheets of heavy cream paper came off Bosqui's press imprinted in glorious color.

Grapes and Grape Vines of California evokes the poetry and romance of the coming of the vine to California. Turning the pages of this elegant folio, one notes how beauty of medium is part of the book's central statement. Here, indeed, is a lovely harvest from the sunny hills of California: the Johannisberg Riesling, from Germany and, before that, from France and Italy, and, even more remotely, from Greece — pale, light, delicate; the Rose Chasselas, tasting faintly of musk; the White Muscat of Alexandria, of North Africa, and Sicily, and the very south of Europe, light green in color, a sturdy, unpretentious grape, for the table or for a decent everyday wine; the heavy Black Hamburgh, flourishing in the fog-cooled Coast Range; the Flame Tokay from Hungary, rank and robust, orange or ruby or rose in tint, dependent upon direct sunlight; the Zinfandel, the most planted vine in California, deep purple, obscure in its origins, perhaps even a totally new California stock (some claimed Haraszthy developed it), fragile of skin, loving the foothills and the highlands, abounding in malic acid, and in its wine, without sugary residue, recalling the raspberry and the strawberry; the seedless Sultana, opulent, from the vineyards of the Near East, running from pale green to ruby, now and then amber-tinted, needing heat for its luxuriance, and protection from fog; and of course the dark and hardy Mission Grape of ancient California lineage, the Ishmael of vines, giving a rough earthy wine, the wine of Spanish soldiers, the wine of Old California.

How thrilling to witness the rebirth of a great book! *Grapes and Grape Vines af California* deserves this new lease on life. It is a bibliographical treasure, most fundamentally, a testimony to the bookmaker's art. But it is so many other things as well: the premier product of a premier printer, Edward Bosqui; a surviving artifact of old San Francisco, where men and women took the time to do things well; and in this translation to Pacific shores of the immemorial arts of viniculture and wine-making, California took one step further in the direction of her assured destiny as a special American place.

An Historical Note on
GRAPES AND GRAPE VINES OF CALIFORNIA

by Leon D. Adams

Grapes and Grape Vines of California is an historical treasure. Produced in 1877 by pioneer color printer Edward Bosqui in San Francisco, it is an excellent early example of fine color printing in the United States, the first ampelography (description of grapevines) published in North America, and a valuable contemporary description of many aspects of California viniculture as it existed more than a century ago. It is also one of the most gorgeous books ever produced in this country; it reproduces with incomparable fidelity all the dazzling colors of grapes, the most beautiful of all fruits.

Hannah Millard's watercolor illustrations of ten grape varieties, painted in the vineyards at vintage time, are contributions to the science of viticulture. They are the first thus far in this country to picture in color both sides of the leaves of grapes, with the clusters and the tendrils on the canes. These are indispensable in identifying individual grape varieties among the more than ten thousand named throughout the world.

The descriptive texts by the California State Vinicultural Association, under whose auspices the book was published, are of additional interest because they represent the prephylloxera era of winegrowing in California. Phylloxera, the aphidlike insect which attacks and kills the delicate roots of *Vitis vinifera* (wine-bearing) vines, was introduced inadvertently to Europe with plant specimens from eastern North America during the 1850s. In three decades it destroyed most of the Old World vineyards, the most destructive plant-disease epidemic of all time. Meanwhile, the pest spread from Europe to California, and between 1880 and 1900 it killed the vines of vast acreages throughout the state. Although first identified on vines at Sonoma in 1873, the phylloxera apparently had not yet done enough damage in the state to be mentioned in *Grapes and Grape Vines of California* when the book went to press. Nor does the text mention the search, then concentrated in the Mississippi Valley, for phylloxera-resistant wild vine roots, onto which most *vinifera* vines subsequently were grafted in both the Old World and California.

A curious historical fact is disclosed by the book's illustration and description of the Catawba grape as one of the varieties then grown in California. In the mid-nineteenth century, California winegrowers made several plantings of Catawba, the pink and purplish grape of the *Vitis labrusca* family, which had made Ohio the leading winegrowing state of the Union. The California acreages of Catawba and of other *labrusca* varieties dwindled in later decades because they failed to ripen with full flavor in the balmy climates of the Golden State. The last remaining Catawba vineyard in California, in the Uvas Valley of Santa Clara County, was removed in 1969. Ohio's leadership among winegrowing states lasted until the 1860s, when it briefly was surpassed by Missouri, which by 1870 was overtaken in turn by California.

While the Bosqui book was in preparation in the early 1870s, published government figures showed only about 36,000 acres of vines in California. Compare this with the 666,000 acres counted by the California Crop and Livestock Reporting Service in 1979. In the 1870s, Sonoma led all other counties, with 6,000 acres, followed by Los Angeles, with 4,335, Lassen, with 3,438, Napa, with 3,238, and Sacramento, with 3,099.

The book's text describes Zinfandel as "at the present time, the vine most extensively planted in California, for wine making"—a rank it retained a century later among the red-wine varieties grown in the state. Before the Zinfandel appeared, the principal California grape was the Mission or Criolla variety, which the Franciscan missionaries brought from Mexico between 1769 and 1779. The unnamed writer of the text was in error, however, in stating that the Zinfandel was introduced from Hungary by the colorful "Colonel" Agoston Haraszthy between 1853 and 1854. Research has shown that Zinfandel was grown in eastern hothouses as early as 1830, before Haraszthy emigrated to America. No such grape has been found in Hungary.

It is interesting that the Bosqui book spells the name of the Johannisberg or White Riesling grape as "Riessling," an early German spelling that I can no longer find used in Germany. This may support claims by descendants of German pioneers of winegrowing in Sonoma Valley that their forebears introduced Riesling to the state directly from Germany. However, it was in a Napa County vineyard that Hannah Millard painted her illustration of "Riessling."

Catawba was not the only grape for which California growers held high hopes in the 1870s, only to abandon it when it failed to show commercial potential in their climates. Another was the third illustration in this book, the pink Rose Chasselas, painted in a Saint Helena vineyard. This is a red-berried form of Chasselas. It is not the Chasselas Doré, most popular table grape of France, which also makes the Gutedel wine of Germany and the Fendant wines of Switzerland. Much better wine and table grapes have since been found or bred for California climates. Current lists of grapes grown in the state no longer mention any Chasselas variety. During the Prohibition era, many vineyards in the North Coast counties of California listed "Golden Chasselas" among their white-wine grapes, and some labeled wines with that name. But the variety they grew was really the Palomino, the grape that, because its juice oxidizes readily, makes most of the sherries of Spain.

The fourth illustration, the White Muscat of Alexandria, was painted in a Santa Cruz mountain vineyard. The text says it is more favored as a table grape than for raisins or for wine. Since Prohibition, however, the Muscat of Alexandria has been classified in California as a raisin variety, although much of the crop, now grown mainly in the San Joaquin Valley, is consumed fresh, and it also makes most of the state's Muscatel wine. For muscat table wines, more delicate varieties are available, such as the Muscat Blanc, also called Canelli or Frontignan.

Muscat Hamburgh, pictured next, is considered by many to be the most beautiful of all grapes. Blue-black or brownish purple with contrasting clear-green tendrils and shoots, it is faithfully reproduced in the illustration Hannah Millard painted in another Sánta Cruz vineyard. Although the text refers to Black Hamburgh as primarily a table grape, it is classed in California among wine varieties. It makes Black Muscat wines in California and the wine called Trollinger in Germany. The gorgeousness of Hannah Millard's Muscat Hamburgh painting inspired the Italian Swiss Colony to buy prints of it from Edward Bosqui & Co. The prints were distributed in the wine trade nationally, and many still hang as ornaments on office walls, including my own. The Colony, however, mistakenly had each print overprinted with a legend that reads: "California Burgundy—Compliments of Italian Swiss Colony, Asti, Cal."

The green Flame Tokay, which Hannah Millard painted in a vineyard at Brighton, near Sacramento, should have been painted in the Lodi district, the only place in California where, during autumn nights cooled by winds and fogs from the nearby Sacramento River delta, this grape develops the flaming-red color that attracts buyers in the fresh-fruit markets of the nation. But Lodi had not yet become the Tokay capital in the 1870s, when the painting was made. The text makes it clear that Flame Tokay has no relationship to the Tokay or Tokaj wines of Hungary, which are produced mostly from the Furmint, a totally different grape.

Sultana, the tiny green or golden raisin grape, was painted in a vineyard near Stockton, in San Joaquin County. It was not until the 1890s that the California raisin industry began moving from the Sacramento Valley to Fresno, which thereafter became the nation's raisin capital. The book was produced before the larger Thompson Seedless grape, also known as Sultanina, became the world's preferred raisin variety. It was in 1872, five years before the book appeared, that the Englishman William Thompson brought from an eastern hothouse to his Yuba County farm the grape called Lady de Coverly, to which he later gave his name. Only 672 acres of Sultana remained in California in 1979, compared to the quarter-million acres of Thompson Seedless, which is valued almost equally for raisins, as a table grape, and for wine and brandy.

Last of the ten grape varieties in the book is the deep-red or reddish-purple Emperor table grape. Hannah Millard painted it in Yolo County, where viticulture declined in subsequent decades but where it began reviving in the 1970s with new plantings of grapes for fine table wines. Despite competition from a dozen newer table-grape varieties introduced since the 1940s, the Emperor, now grown entirely in the San Joaquin Valley, continues to lead them all, with 20,812 acres counted in the 1979 survey.

The printing establishment of Edward Bosqui & Co. was destroyed in the 1906 San Francisco earthquake and fire, and with it Bosqui's stock of *Grapes and Grape Vines of California*. Nine copies are known to exist.

GRAPES AND GRAPE VINES

OF

CALIFORNIA

PUBLISHED UNDER THE AUSPICES OF THE

CALIFORNIA STATE VINICULTURAL ASSOCIATION

OLEOGRAPHED BY WM. HARRING

FROM ORIGINAL WATER COLOR DRAWINGS BY MISS HANNAH MILLARD

SAN FRANCISCO

EDWARD BOSQUI & CO.

1877

PREFACE

DURING the last quarter-century, some elaborate and costly works have been published on the grapes and wines of the chief vine-growing countries of Continental Europe. Of these, several not only had a utilitarian value for those engaged in the culture of the vine, but were æsthetically interesting to a much larger class for the beauty and excellence of their illustrations. Some of these works were strictly local, being merely monographs, each relating to a single viticultural district. Others, such as Victor Rendu's "Ampélographie Française," and the Ampélographie Rhénane," were devoted to the vines of an entire country or region.

The magnitude of the vinicultural interest in this State, the extent of land already covered by our vineyards, and the still vaster area peculiarly adapted to the cultivation of the grape, together with the large amount of capital so invested, have seemed to amply warrant the present attempt to do for California what has so often been done in Europe—sometimes at great expense—for vine-growing regions of far inferior prospective or even present importance.

This Work, published under the auspices of the California State Vinicultural Association, is designed to accomplish the same general objects aimed at in the French and German "Ampélographies," together with some others of a special character, which will be readily appreciated by California vine-growers. Though all the varieties of grape, cultivated in our vineyards, belong to foreign stocks which have been described with exactness, some of them are, from various causes, undergoing here changes that tend to impair the value of those descriptions as a means of identifying California vines. True, the modifications produced in time by a new soil and climate, are exceedingly slow in their development and have not as yet seriously affected the distinctive characteristics of the vine varieties introduced into this State,—still superficial changes have been produced in their general appearances, in the size and flavor of the fruit, etc., sufficient to perplex the non-botanic observer; while the more radical and embarrassing modifications resulting from the production

of hybrids, have contributed much to the confusion existing in some of the California vine-classifications. It is believed that the present work will in some measure remedy this confusion, showing as it does, by means of its accurate and life-like pictures—and more clearly than could the most elaborate verbal description—the distinctive traits of the principal varieties cultivated in this State.

Another incentive to the preparation of the present work, was a desire to exhibit to the people of the East and of Europe, the splendor and variety of the products of the California vineyards. And stimulating this natural pride of country and State, was the thought that a deep and wide-spread interest, to the benefit of all, in the grape and wine future of California, would be awakened by giving to the world adequate representations of her best fruit products.

To insure the utmost accuracy and truth to nature, the artist, Miss Hannah Millard, visited at vintage season the vineyards in various sections of the State, and made on the spot paintings for this volume, each depicting an entire selected cluster, a portion of the wood, and two leaves, showing the upper and the under side. That the illustrations should not convey false or exaggerated impressions, only average clusters were taken for delineation. The drawings have been executed most conscientiously, being in every instance, exact copies; and fac-similes in form, color and size, have been made of these on stone by the oleographist, Mr. Harring, with a patience, labor, and skill appreciable only by an artist. This costly and arduous work seemed indispensable, in view of the fact before stated, as also that many vines when transplanted to California, produce fruit much larger, both in bunch and berry, than they usually do in their parent soil.

In fine, it may be safely affirmed that "THE GRAPES AND GRAPE VINES OF CALIFORNIA" will be found fully equal to the most costly works of a similar character produced in Europe: nothing having been spared to secure this result, the work is necessarily expensive, and could not well have been made less so, except by such a lowering of its designed character as would have rendered it unworthy of comparison with the best similar French and German publications.

THE MISSION GRAPE VINE.

IN a Work descriptive of the "Grapes and Grape Vines of California," precedence is naturally given to the Mission variety, as it is the only one of the transplants to the Pacific slope whose European lineage has been apparently lost or effaced in its American naturalization. This fact, more perhaps than its intrinsic value, must make the vine always interesting no less to the people of this State, than to the botanist. Now regarded as indigenous solely to our soil, it has latterly been familiarly called "The California." Though the early Spanish and Mexican settlers found growing wild in this State a vine of the species Labrusca, from which a tolerably good wine and brandy were made, this was soon abandoned for the "Mission" vine, which they found growing in the vineyards of the Catholic Missions established along the Coast from San Diego to San Francisco. Hence the name. That it was brought by the priests to this State from either Spain or Mexico is pretty certain, but, though good judges believe it to be identical with the Val de Peñas of Spain, the transplantation seems to have effected such a change in its general appearance, that its parentage can no longer with certainty be assigned.

The appearance of the Mission vine is not uniform in all parts of California. In the San Joaquin Valley and some of the mountain districts, it is in every respect a most luxuriant vine, while in Sonoma and other localities it is less rank, the leaves, clusters and berries being of a diminished size. So marked indeed are these changes, that some consider them as representing different varieties of this vine, while others regard them as the result of merely diverse circumstances of growth.

The Canes are usually large and long, with a moderate distance between the nodes.

The Leaves are of a subdued green color, showing a smooth and shining upper surface. They have two sinuses, making three principal divisions: these sinuses extend to about half the length of the lateral large veins or ribs. The side lobes are also cleft, but not deeply.

The Petioles are of moderate length and have a reddish purple tint.

The Grape-Clusters of this vine are large, its berries growing from numerous long pedicels attached to a long peduncle,—the whole seeming to be loosely put together. The form of the bunch is usually a long graceful pendant, broad at the top and quite narrow at the terminal berries. Though not compact, still the clusters are well filled with berries. These are round and purple, varying in size according to the locality. They have a thick bloom and their juice is abundant, sweet and agreeable.

The Fruit-stems are of a whitish green color, and somewhat bleached when the fruit is fully ripe.

A loose, porous soil, sandy, gravelly or loamy, is well adapted to the Mission; it will grow well and give a fair yield in a soil too poor for cereals or other staples. In the hot interior valleys of California, it is grown with particular advantage, and with nearly equal success in many other locations. It is pruned from December to the middle of March, two, three or four eyes being generally left upon the wood.

It is less subject to mildew (other vine-diseases being as yet little known here) than are some other varieties, and is considered one of the most productive vines of this State. It is said that the records at one of the Missions show that, during a period of eighty years, there was not even a partial crop failure of this productive, vigorous vine. Of late, the vine-growers have grafted, to some extent, foreign varieties upon its healthy, hardy stalk.

The Mission is deservedly prized and in great demand as a table-fruit, and, as such, is deemed superior to the grapes grown in the open air on the Atlantic side of the Rocky Mountains. Hardy, strong and prolific, it will be set down, by Europeans at least, as the Ishmael of the vine-race.

The chemical constituents of the juice of the ripe berry, were given by Dr. Blake to the California Academy of Sciences, as follows: Sugar, 21.5; Free Acid, .6; Malic Acid, .11; and its specific gravity a little more than that of water, or 1088. From his careful analysis it appears that this grape contains a remarkably large percentage of sugar, considerably more than does the Riessling, the sweetest of three foreign varieties tested by him. The usually strong and heady character of the wine manufactured by ordinary modes from this grape, depends no doubt on this abundance of the alcoholizing material. Hence this grape would seem well suited for the production of sweet wine or brandy, to which uses it has been put in this State. The predominance of this element makes it valuable also for blending with European wines of a lighter character, needing more spirit. The small proportion of Malic Acid (which gives the bouquet to wines) found in its juice, is very noticeable. The Riessling contains .57, or nearly six times the quantity. A peculiar earthy taste is usually found in the wine, which some vintners think is an essential property of the berry, while others strongly dissent from this notion. Unanimity on such points cannot be expected, until the viticulture and viniculture of this State shall have passed out of their infancy, and the experience of a few more years shall have been gained. Though the dissatisfaction of wine-makers with the Mission, as a wine grape, be not without foundation, still its characteristics may, in the end, prove it to be very valuable if not as a principal, yet as an auxiliary, in the improvement of wines made from French and German varieties. If used in judicious proportions, it can give these greater alcoholic strength, greater body and a sweeter and more mellow taste, without altering their bouquet. The White wine made from this grape is remarkable for its hardiness or preservative quality; it bears shipment to any climate even when very young, and resists careless handling to an uncommon degree.

The cluster of the Mission grape represented in the illustration, was taken at a vineyard in the Valley of Sonoma.

THE JOHANNISBERG RIESSLING.

IN the vineyards which cover the hill-sides and slopes of the Rhine, the Riessling has for ages maintained a great reputation.

Although it is scarcely at all cultivated in either France or Italy (the climate or soil being adverse), it must be well appreciated by the French, who have named it "Gentil Aromatique," and "Noble Aromatique,"—terms strongly expressive of the fine and inimitable aroma which the berry imparts to the wine.

The Riessling, it appears, came originally from Greece, and was introduced, through Italy and France, into the Rheingau. Of the California transplants, this is the vine the most deservedly prized by wine-makers, its fiuit giving a liquid as copious in quantity as excellent in quality. Its acclimation here has increased the quantity of its yield, through the superior size of both cluster and berry—this luxuriant growth not at all impairing that delicacy of flavor which is its most precious characteristic. This soil and climate seem, in fact, to be highly congenial to it, since it escapes here not only the cold winds and early frosts from which it often suffers in some districts of Europe, but also the extreme heat to which it is very sensitive, and which prevents it from thriving in Spain, Italy, and Southern France.

In California, this vine flourishes in nearly every locality suitable for the grape,—as well in the valleys and plains as on the slopes of the foothills. In these situations a crop is almost invariably sure, while in Europe, the grape being slow to ripen, much uncertainty attends its culture, unless a favorable location protect it from early frosts and cold winds. All soils are adapted to this grape, except the heavy adobe and one that is wet, hard or cold. The best results are obtained in ground humid without excessive moisture, and in one formed of decomposed granite, clay-slate, or a reddish clay, any of which, mixed with sandstone and loam, makes a good soil. It thrives in sandy and gravelly ground, provided this is light, or worked to a depth that will allow the fibres of the root to penetrate to the moisture. In the celebrated Johannisberg vineyard, where the Riessling is the principal vine, the soil is composed of variously colored stratified marl.

It is maintained by some, that there are two or more varieties of the White Riessling, but perhaps the differences noted in the size of the plant, in the leaf and fruit, in the color of the wood and the flavor of the berry, are attributable to differences of soil, and circumstances of growth. The Red and the Black Riessling are considered to be distinct varieties These are not extensively cultivated in Europe for wine purposes, though the latter yields a good red wine. The "Franken Riessling" probably belongs to the "Sylvaner" or "Round Leaf" family.

The wine made from the Riessling requires age to develop its choice aroma and flavor; it is then dry, has a delicate flavor, and much less body than have other California wines.

An accurate chemical analysis by a resident expert, shows the fruit of this vine to contain,—of sugar, nearly 19 per cent; of free acid, a little over 1 per cent; and of Malic acid, nearly 6/10 of one per cent.

The Riessling is a vigorous though not a rank plant, the stems, canes, leaves, clusters and berries being small. The stock or trunk of the vine has a bark of a darker color than that of other species of grape vines.

The Canes are of moderate length, making small, close-jointed wood, of a light yellowish color while the plant is in leaf, but of a silvery or pale grayish shade after the fall of the leaf.

The Leaves are small, being but little larger than those of the Chasselas; they are deeply cleft into three or five irregular-looking lobes, approximating, however, sometimes, when without sinuses, to a circular form. Their surfaces vary, being sometimes smooth, and sometimes furrowed by the veining. The upper surface is of a deep glossy green, turning yellow as the fruit ripens; the under surface is of a pale gray green, sometimes cottony.

The Petioles are short and of a light green, and sometimes of a light purple color. The Pedicels are long for the size of the berry, from which they readily separate when the fruit is ripe, the peduncle then becoming very brittle.

The Clusters are small and never more than of a medium size; they are close and compact, but not crowded with berries; sometimes they have shoulders, less frequently a side bunch; but the ordinary type is an oblong cluster without shoulders.

The Berries are round and sometimes a little oblong; they are small, or of a medium size, and of a pale green color, with a heavy bloom before ripening. When perfectly ripe, they acquire yellowish and flesh-colored tints, and are marked with minute dark-brown spots, more numerous if the vines have grown on poor soil. In most of the California vineyards where the Riessling is grown, the grapes are but slightly marked, many clusters being found in which the berry exhibits no spots. The skin of the berry is thick but soft; the flesh is pulpy, sweetish and aromatic.

The Riessling readily takes root from cuttings, and in this State is found to bear much better with long pruning; here it is usually pruned with two long canes, having eight eyes to each, and from four to six short spurs, with not more than two or three eyes; the vine is trained to stakes four or five feet high.

The best proof that the climate and soil of this country are admirably adapted for a perfect development of this sensitive and fastidious vine, will be found in a comparison of the analysis of the juice made by Fresenius, in Germany, with that made here by Prof. Blake. The former shows the quantity of sugar as 15 per cent and of free acid as .53 per cent; the latter gives 18 per cent of sugar (a valuable increase in an acid grape like the Riessling), and 1.1 per cent or twice as much of free acid. If on the presence of the last depend (as is probable) the quality of the wine, it is evident, as Dr. Blake remarks, that the Riessling may give us here, when grown in suitable localities, a wine even superior to that it yields in Germany. A later analysis in 1872, by Messrs. Thudicum & Dupré (authors of an exhaustive work on European Viticulture), gives the interesting fact that the chemical constituents of the German growth, as compared with those of the Californian, would indicate a superior growth for this vine on the Western Continent. It gives 16.2 per cent of sugar, as against 18.7 here; 1.33 of free acid, to 1.1 here; and of *Malic* (called by them "combined") acid, .196 per cent to .57 here, the most valuable difference of all in favor of this State.

These figures given by most reliable men, ought to speak more eloquently to the world of its vine future than could an elaborate treatise without them. If the splendid fruit products and wine industries of the Old World perish, provision seems already made that they shall survive for mankind in the New.

The specimen of the Riessling given in the illustration, was taken at a vineyard in Napa Valley.

THE ROSE CHASSELAS.

ALL species of the Chasselas vine have long enjoyed a high European reputation. In France they are prized for table-grape above all varieties. French writers call this genus, "Doucet;" the Germans name it "Gutedel."

Many of its numerous varieties closely approximate to each other in grape color; some are a pale whitish green, varied with a faint tinge of rose, yellow, gold or purple; while others are a deep ruby, red, violet, or purple. That selected to illustrate the "Rose Chasselas," is one of the most highly esteemed, and is extensively cultivated in California. By some vineyardists it is called the "Victoria," by others, the "Red Chasselas."

The following are some of the principal varieties of the Chasselas, described by French and German writers.

FRENCH.	GERMAN.
Le Chasselas Blanc - -	Weisser Gutedel.
Le Chasselas Jaune - - -	Gelber Gutedel.
Le Chasselas Petit - -	Kleiner Gutedel.
Le Chasselas Blanc Hâtif - -	Früher Weiser Gutedel.
Le Chasselas Croquant - -	Krachgutedel.
Le Chasselas Rouge - -	Rother Gutedel.
Le Chasselas Royal - - -	Fruhrother Gutedel.
Le Chasselas de Fontainebleau.	
Le Chasselas Musquié	

Of the above, some are sweet, some have a sub-acid taste, and others, a flavor slightly suggestive of musk.

The Chasselas vines are in marked contrast with the large and vigorous growths of the vineyard; theirs is not a robust structure, the leaves and canes being small.

Though resembling the others in the general appearance of its foliage, the subject of our description is not so *petit* as are some other members of the family.

The Main-stock, or Trunk, of the Rose Chasselas has an adherent bark.

The Canes are of medium length, with short internodes.

The Leaves are small and elegant in form; they are deeply cut into five divisions by the sinuses, the largest of which passes through the mid-vein. Their color is a medium green, until the fruit approaches maturity; then it assumes a rich yellowish green. The veins are nearly white. The young leaves of all the Chasselas vines have a reddish tint; this, their diminutiveness, their deep, well-defined cuts, elegant form, and length of tendril, constitute peculiarities by which the Chasselas vines are readily distinguished.

The Peteoles are pink or purplish.

The Clusters of the Rose Chasselas are large (sometimes weighing five and a half pounds), and are well filled though not crowded with berries. They have two or more broad shoulders over-hanging the main pendant, which is long and slender at the terminal berries.

The Berries are round or nearly so, and not large; the skin is rather thin, and of a deep, rich ruby color, overspread with a delicate, light-purple bloom. The pulp is fleshy, sweet, slightly aromatic, and very agreeable.

The Rose Chasselas is easily grown from cuttings, and usually throws out suckers.

The fruit of this vine ripens early, and is not easily injured by transportation—facts that account for its great demand.

The specimen illustrated was grown at St. Helena, Napa Valley, where the soil is a loose loam, and one admirably adapted to this vine. In most Californian vineyards, especially those in warm and dry locations, the Chasselas vines are trained low, and are not allowed any support; but where the ground is very damp, or liable to be visited by frost, it is necessary to stake them and keep the clusters from the ground.

There are few localities, regarded as suitable for viticulture, in the valleys, or on the hill-sides of this State, where, with proper care, the Chasselas vine will not succeed.

THE WHITE MUSCAT OF ALEXANDRIA.

THIS highly esteemed vine is probably a native of northern Africa, or of the adjacent parts of Asia, whence it was introduced into southern Europe. Though known among the vine-growers of California as the "White Muscat of Alexandria," it figures in the catalogues and vinicultural treatises under a great variety of names, among which are the "Muscat Damascene," "Jerusalem Muscat," the "Frontignac of Alexandria," the "Tottenham Park Muscat," and the "White Muscat." Its merits have been widely appreciated, and its cultivation has been attempted in nearly all the grape-growing countries of Europe. These trials have met with success in Spain, in the southern parts of France and Italy and in the countries along the Mediterranean, where the climate most nearly approximates to the fervor of the torrid zone. It flourishes in the vineyards of Languedoc and Provence, and finds a still more congenial home in the valleys and hill-sides of Corsica, and the deep volcanic soil of Sicily, enriched by the debris washed down from the steeps of Mount Etna. In the valley of the Rhine the climate appears to be not sufficiently tropical. There it succeeds only in the most favorable and well-sheltered situations.

In California it has prospered in a great variety of situations ; in the smaller valleys of the Coast Range, and in the larger ones of the interior ; on the plains and hillsides and on high elevations. The specimen represented in the chromo was taken at a vineyard situated in the Santa Cruz Mountains, on the eastern side of the ridge between San José and Santa Cruz, at an altitude of more than seventeen hundred feet above the sea, where the soil is a loam, mixed with gravel, clay and a little limestone.

The Stock of the Muscat of Alexandria is tolerably robust and vigorous. The Canes make a long growth, while the distance between the joints is short.

The Leaves are not large. They are of a light-green color, with a smooth upper surface, and are ordinarily marked off into five parts. The incisions, forming the lobes, are of moderate depth. The Petioles are of a pink color.

The Clusters have strictly no prevailing type of form, being generally much branched; they seem loosely put together, and are very rarely compact. They are sometimes long and slender, sometimes small and nearly round ; sometimes with shoulders, sometimes with a side bunch, and again, without either.

The Berries are usually large, often very large, lacking uniformity either in size or shape. Large, medium and small berries are found in the same cluster. Their most usual form is ovoid, with the smaller end attached to a long pedicil; but they vary much from this type, and are often nearly round. In color they are generally a pale green, though occasionally exhibiting, at different stages of maturity, amber, golden, or purplish tints. They are firm-fleshed, have large seeds and thick skins, and, while very sweet, have a rich, musky flavor, similar to that of the White Frontignan, the Royal Muscadine, or any of the Muscat family.

Though the White Muscat has many of the requisites of a good raisin-grape, its large seeds and thick skin constitute serious drawbacks for such a use. Its aromatic and musky flavor makes it a favorite table-grape and ensures it a high market price. This also gives it a certain value in wine making, as a flavorer of the juice of other varieties.

This vine is but little subject to mildew or other disease, though its blossom is sensitive and sometimes suffers from blight. It is a heavy bearer, especially in the mountains, or in any location, high or low, that is not far from the ocean. It gives two crops, of which, if the first prove light, the second is generally abundant. In pruning, the vine is usually left with two or three eyes.

THE BLACK HAMBURGH.

THE Black Hamburgh is a widely known representative of the "Large Race" grape-vines, called by the French and Germans, "Trollinger." In Europe, this vine has many appellations, being called very commonly in Germany "Frankenthaler," and also "Malvasier," a name common with other species. Blue Tyrol" is one of the titles it bears in Southern Europe; "Black Chasselas" is another of its names, although its traits are not at all identical with those of the Chasselas family. In California its usual title is "Black Hamburgh."

Besides the black, there are also red and white varieties of this vine, but the two latter do not possess the economic value of the former, and are not extensively cultivated. In all countries the Black Hamburgh is chiefly valued as a table grape, and has proven in California a most estimable grape for this purpose, bearing generously, and readily adapting itself to all vine-growing localities.

Its growth is medium, though in every respect larger than the Chasselas race.

The Main stock has a rough, much-furrowed bark.

The Canes are somewhat large and long, and are slightly arched at the nodes, the spaces between the nodes being from medium to short.

The Tendrils are large, having many branches.

The Leaves, which are large, are usually broader than long, and are divided into from three to five lobes by sinuses, which are much cut or rounded out. They have mostly a smooth upper surface, although frequently one is seen much furrowed by the veining; their under surface is slightly cottony. The general hue of the foliage is a deep, warm green, becoming quite yellow when the fruit is ripe.

The Petioles are long, of a light crimson color, which tint is also carried into the large veins of the leaves they support.

The Clusters vary from large to small size, and are usually shouldered; often many shoulders hang around the main bunch, giving it a pyramidal form, though sometimes they are found of a quite different outline. They are very compact with berries.

The Berries, which are large, approach the globular form; generally they are slightly longer than wide, and are egg-shaped, and sometimes oblong. They have a sweet, firm pulp, and the skins of those grown in the mountains have a flavor which leaves in the mouth a peculiar after-taste. Their color is an intense brown purple, almost black, over which is spread a bluish bloom.

The Peduncle is large and long.

The Black Hamburgh is successfully propagated from cuttings. The Coast-range Mountains are better adapted to its culture than the warmer valleys, in the latter it is liable to sun-blast. In the mountains it is pruned leaving six spurs, and two or three buds upon each spur.

A soil which contains much iron as well as chalk, is suited to it. The ripe grape clusters contain .13 per cent. of malic acid.

The specimen illustrated was taken from a vineyard located on one of the declivities on the Western slope of Mount Bache, in the Santa Cruz mountains, a locality having an elevation of seventeen hundred feet above the sea-level. That favored section is protected from the North winds by the broad, high summit of that mountain, and from the ocean fogs by other peaks nearly as broad and high. There, as well as in other cañons in those mountains, many varieties of table grapes attain rare excellence. They slowly ripen in the mild, equable climate, attaining perfection in October and November,—too late for curing them into raisins before the Autumn rains commence, consequently the crop is mostly sent to market for the table.

THE FLAME TOKAY.

THE attractive grape known in California as the Flame Tokay, is much valued for its beauty, its liberal yield, and excellent keeping qualities.

Its name indicates an Hungarian origin, though it is not the grape from which the renowned imperial Tokay wine of Hungary is made. It has other names, among which are "Red Grape of Taurida," "Red Lombardy," "Wantage," and "Rhenish Red." From these titles it may be inferred that this vine has flourished in Southern and Central Europe, to the countries of which it would seem better adapted by its strong growth and luxuriant habits, than to colder Northern regions. In California, however, it has attained its best growth in well protected localities, at an altitude of seventeen or eighteen hundred feet;—a statement which may be questioned by those persons who have succeeded admirably in cultivating it in the warmer valleys, remote from the sea.

As its rank, robust appearance indicates, the "Flame Tokay" is a strong grower and heavy bearer.

The Trunk has a rough, much furrowed bark.

The Canes are long, but not so large as their massive clusters would seem to require. At the nodes they are of a purplish hue; the inter-nodes are of medium length.

The Leaves, which are mostly large and rough, are generally divided into five parts, the central division being rather long. They are deeply notched; small teeth alternating with the larger. In color they are of a medium or dark green, from which they change in Autumn to every shade of russet, occasionally varied with the brightest crimson and scarlet. The Petioles supporting them are usually long.

The Clusters are large, sometimes weighing eight pounds. They have a pyramidal form, with broad shoulders, and are usually well filled with large, roundish, irregularly-shaped berries. Their aspect, as they hang upon the vine, is not that of a graceful pendant, but of a solid, massive bunch.

The Berries are far from being uniform and symmetrical in shape; generally they are ovoid;—often round, or nearly so. Their tendency is to be broader at the end attached to the pedicils, having the smaller end a little flattened, and their sides somewhat indented. An umbilic is apparent. Each berry takes the orange, ruby or rose tint, as the light shines upon or through it. These, colors when the berries are ripening, seem superimposed upon an undertone of green, but when ripe the green disappears in those localities where the fruit colors well.

The flesh or pulp is coarse-fibered, but firm and when fully ripe is sweet and watery, with some flavor; the skin is thick. The fruit-stems are light green, occasionally tinted with rose.

The Flame Tokay ripens in September and October. In some seasons and localities the bunches color poorly, which failure of color is owing to an over-crop. The vine is pruned in February, with three eyes to each spur.

It is not a good raisin grape, but is vlaued because the thick skin of the berries enables it to bear transportation well, and keep in good condition for many weeks, while the great beauty of the clusters insures it a ready sale.

To secure the best quality of fruit, the vines should be closely pruned. They are not staked, but are left to grow with their superb clusters near the ground.

The specimen illustrated was taken from a vineyard in the town of Brighton, near Sacramento; the soil of that locality is a rich alluvium, well adapted to this vine. In the mountains, where the soil is less rich and deep than in the Sacramento valley, it is equally productive.

THE ZINFANDEL.

AT the present time, the vine most extensively planted in California, for wine making is the Zinfandel. It is more productive than the Mission, without its alleged defects, and possesses fine and excellent qualities, which are rarely found in connection with so great productiveness. Being easily cultivated in most grape-growing localities, it is expected in the future to prove the standard red-wine grape of this State. Its prolific first crop is early harvested, and is succeeded by an abundant second crop, which amply repays the trouble of gathering.

The thin skin of the berries cracks badly, under the treatment necessary for transportation to market, consequently it is not a good market grape.

The Zinfandel vine was first brought to this State by the late Col. Agoston Haraszthy, between the years 1853 and 1854. It was imported by that gentleman from Hungary, his native land, where he had known it to be held in great esteem as a wine grape, and was first planted in his vineyard at Crystal Springs, San Mateo County, and afterwards in Sonoma, in 1857 and 1858, whence it was distributed throughout the State. The first few gallons of wine were made from it in 1862, after which date its reputation as a wine-making grape spread very rapidly throughout the entire State, and competent authorities soon began to rely upon it as the vine which in the future would produce those fine red wines usually denominated clarets.

In whatever locality of this State the Zinfandel is grown, the wine is always good, but experience has shown that its peculiar, fine and distinctive characteristics are enhanced or diminished by the kind of climate as well as quality of soil in which the plant is reared.

The Zinfandel is a vine of medium vigor; neither trunk nor canes are large for so heavy a bearer.

The Canes are of medium length; light-colored, with the bark purplish at the nodes. The internodes are short, or of medium length.

The Leaves, which have a great variety of shape and appearance, sometimes approach the round form, being slightly cut out by the sinuses; again the sinuses are deep and much rounded out, leaving three or five lobes, with the division through which the mid-vein passes quite long. The texture of the leaf is thick, and the color dark-green upon the upper surface, and whitish green on the underside, owing to the down with which it is covered. Both surfaces are much marked by the veins, and often crimped and furrowed as well as ruffled.

The Cluster has a very unusual form: from the main bunch is projected a side bunch or shoulder (sometimes more than one), often nearly as large as the principal pendant, with which it forms a very open angle. Generally the clusters are large and very compact. The bunches of the second crop are quite unlike those of the first, being much smaller, loosely put together, less solid and without the formal, rigid outline of those earlier ripened, and the berries lighter colored.

The Berries are large, or from large to medium size, round, dark purple, with a heavy bloom. When ripe, the pulp is sub-acid and very juicy. The skin is thin, breaking easily.

THE ZINFANDEL.

The foliage of the Zinfandel is not abundant, its numerous clusters being, in consequence, much exposed to the sun. To avoid sun-blast in localities where the heat is excessive, with little moisture in the soil, the vine may be pruned, leaving many long canes, which, in the spring, should be brought together and tied at some distance above the trunk, so that the whole will grow in a round, compact mass, the leaves overhanging and protecting the bunches. In climates more temperate, shorter pruning, or the ordinary spur manner, is found to answer its requirements. In favorable localities, it is not thought to be more susceptible to mildew or sun-blast than any other vine. The foot-hills and high lands are the locations where it gives its most prodigal yield, although it is always a good bearer in any vine-growing section. A soil containing much iron is well adapted to it. In the Sierras, where this vine succeeds admirably, the soil is largely made up of decomposed granite. The specimen here illustrated was taken from a vineyard in Pleasant Valley, Solano County. The soil of that locality is a rich alluvium, and the climate very warm during the day and night. In that deep soil and unusually warm climate, the clusters of the Zinfandel grow to their largest proportions, probably larger than in most sections of the State, and are ready for the vintage by the twelfth of August, three or four weeks earlier than they are harvested in other places.

The Zinfandel berries abound in malic acid, and have a smaller proportion of sugar than most wine grapes, according to the analysis made by Professor Blake. If the bouquet of wines is mainly owing to the presence of malic acid, then the clusters of the Zinfandel have a great advantage over those of most other vines. They possess a larger proportion even than the Riessling, which is famous for its abundant bouquet.

When the bunches of the Zinfandel are gathered at an early period of ripeness, a light, agreeable, dry wine may be made, having a pleasant tartness and a rich flavor, containing the combined qualities of the raspberry and strawberry, to which is added a perfume resembling the violet. These qualities are apparent in both the red and white wine.

When the fruit is left to become very ripe and consequently sweeter, the red wine resembles port and the white wine is similar to sherry. The red and the white wine are both in great demand, and sell for the best of prices ; and both compare well with the most renowned of foreign wines.

THE SULTANA.

THE vine bearing a small, seedless grape known as the Sultana, is an Oriental,—as its name indicates—flourishing in the distant East, as far as Persia and Afghanistan. It is most productive and profitable in those countries of the Orient which border upon the Mediterranean. Traffic in its raisins is an important item of the commerce of Asia Minor. To some extent, it is cultivated in Southern Europe,—Spain, France, Italy, Greece and the islands of the Archipelago.

It was introduced into the United States by the agricultural branch of the Patent Office, in the year 1853, and distributed by that department among those States thought favorable to its culture.

Until within a few years, grapes profitable for wine making and the table have mostly claimed the attention of vine-growers in California. Recently, vineyards have been planted for the production of raisins, and vines are now sought adapted to that purpose. The Sultana, though not yet extensively grown, is highly esteemed by those cultivators who have planted it and proven its worth. It possesses all the requisites of a good raisin grape, except large size, and has the advantage of ripening two weeks earlier than the Muscat of Alexandria and Muscatella, and also of drying in from four to six days, while the former requires more than double that time. The clusters and berries attain in California a fine development for grapes which are classed in other countries among the pigmies of the vine tribes.

In many respects, the Sultana resembles the Corinth vines, the latter consisting of several varieties of small, seedless grapes; but there are some differences which separate it from that family. The leaves of the Corinth are darker of color and thicker of texture than those of the Sultana; their undersides are thickly tomentose, which is not peculiar to the other. Usually the clusters of the White Corinths are oblong and destitute of shoulders, while the pendants resembling shoulders, are thickly crowded around the main bunch of the Sultana. The general appearance of the plant and size of clusters and berries place the latter between the Chasselas and Corinth vines.

The Sultana vine is a vigorous but not rampant grower. Main stock and canes are large for a vine bearing so small berries.

The Canes of the Sultana are of a light yellow color; the internodes from medium to short.

The Leaves are small, and not deeply cleft. Usually they have three principal divisions, and are mostly smooth, without crimps or furrows, and destitute of nap upon their under side. Their color is light green.

The Petioles which support them are usually white; when colored, of a light purple tint, and are short.

The Clusters have a peculiar outline, and vary but little from one type. The central bunch is long and slender; around it hang many shorter bunches, which droop about its column, rather than project from it like shoulders. The whole group—(for it is a group of bunches, closely set with pearl-like berries)—is very graceful and elegant.

The Berries—which are smaller than those of the Chasselas vines—are round, or nearly so, and when fully ripe, where exposed to the sun, acquire various colors. The shaded parts of the cluster remain a pale green, covered with a delicate bloom, while the more exposed berries become of a rose or ruby hue; or, transmitting the light, golden and amber-tinted. To this variety is sometimes added faint suggestions of purple. They have a pleasant, sweet taste, no seeds, a thin skin, and are rather firm-fleshed.

The Sultana is easily propagated, and has not shown itself susceptible to mildew or other disease in those localities in California where it has been most cultivated.

Pruning with short spurs is considered the best method.

"Kishmish" is one of the names by which it is known in the Orient.

When the Sultana is planted in dry soil, it is said to succeed better than the Muscat of Alexandria in the same location. Situations exposed to cold winds and fogs are not favorable to it. The foot-hills and many valleys of California afford ample opportunity for the culture of this excellent vine, whose raisins are so valuable for dessert and culinary purposes

The specimen represented was taken from a vineyard near Stockton. The soil of that locality is a deep alluvium, the climate dry and warm, affording a good opportunity for converting the clusters of the Sultana into raisins.

THE CATAWBA.

THE Catawba is a well-known vine in the States east of the Rocky Mountains. A hardy plant, it is well fitted to endure the vigorous winters of the Atlantic and other States remote from the Pacific Coast, where it is one of the standard grapes for wine-making, especially in Ohio and the Lake Shore region.

The comprehensive *Genus Vitis*, among other species includes two of great note: Vinifera or vines of the Eastern Hemisphere; and Labrusca, one of the species of wild vines native to North America. The Catawba is a representative specimen of the last mentioned: though not yet so ameliorated from its native wildness as to possess the high-bred virtues of those European vines which for ages have been subject to the care of man, still the Catawba has many sterling qualities, and is so productive and hardy that it is highly esteemed in the Atlantic States.

Its native woods and groves are those of North Carolina and the States contiguous. There it climbs over and covers trees of the largest growth.

It is nearly allied to the Isabella, and like it, is a strong, robust grower, with large leaves and long canes.

The Main-stock is large, with rough bark.

The Canes are long, with long internodes.

The Leaves are nearly heart-shaped, large, and not usually lobed, thick in texture, with a whitish, cottony under-side, and bright shining upper surface. In autumn these vines are the gayest in the vineyard. They contrast with their bright or pale greens, the tones and semitones of red and yellow, purple, brown and maroon in the most daring and dashing manner.

The Clusters are of medium size, never very large, loose in structure, and slightly shouldered, somewhat resembling the Mission, though not so long or large, but more compact.

The Berries are round, of medium size: their color ranging from pink to maroon or purplish red; sometimes they are nearly as dark as the Isabella. Their skin has a strong pungent flavor. The berries of the Catawba share with the other wild vines of America the traits which help to distinguish them from the Vinifera of Europe, which traits are: seeds firmly imbedded in a hard pulp, and which slips easily from the skin.

The Catawba fruit is much improved by cultivation, becoming less acrid and acid; clusters and berries become larger and in every way finer. It is a good market grape. The vine is easily propagated from cuttings, is very prolific, not subject to mildew or other disease In the mountains, when the vines are old, they are pruned, leaving ten canes with four or five buds to each.

The illustration represents a specimen grown in El Dorado County. There the soil to which it is well adapted, is composed of decomposed granite; a sandy loam is unsuited to it.

East of the Rocky Mountains the Catawba, like other vines, is raised upon a trellis; this expense is unnecessary in California, as it will furnish finer fruit when pruned, staked and kept near the ground.

It has been introduced and cultivated with some success in Malaga and other parts of Europe, but probably develops its best fruit in California; the long and dry summers favor its complete ripeness, when it is quite sweet, which is not always the case in the East.

Catawba wine has very much the flavor of the fruit, and persons who enjoy the pungent, somewhat astringent taste of the grapes, generally like the wine.

THE EMPEROR.

THE Emperor, is a name given in California to a foreign vine grown in the thrifty vineyards of Yolo County, and to some extent in other parts of the State. It merits this regal title, by reason of the unsurpassed size and beauty of its clusters, and other qualities valuable in a market table grape.

It is a rank, vigorous grower. Around a Main-stock of medium stoutness are thrown out Spurs and Canes of ordinary size, interspersed with others of very unusual dimensions: their color is yellowish brown, with pinkish bloom at the nodes.

The Internodes are from short to medium and quite long.

The Tendrils are unusually large and strong.

The Leaves, which are larger than those of any other grape-vine, are not deeply lobed, neither is their margin deeply serrated; small teeth alternate with larger ones. Their upper surface is of ordinary smoothness, but sometimes roughened by the veining, lustrous, and of a bright green color; the under surface, pale grayish green, owing to the fuzz or down with which it is thickly covered.

The Petioles are long, often very long, of a purplish color.

The Cluster is ordinarily a large, long pendant, hanging from a peduncle also long and much branched, broad at the top without over-hanging shoulders, having many long, rambling fruit stems, which with their fruit form a graceful tapering bunch, very small at the terminal berries; their whole mass so pliant and yielding, it may easily be rolled and doubled up without danger of breaking.

The Berries are mostly oblong, very large and long, sometimes round or approaching roundness; their color deep red or reddish purple, and are covered with a heavy purple bloom. The pulp is firm and has an agreeable flavor.

A rich sandy loam containing much iron and lime is the soil in which the Emperor thrives best. It is not subject to disease, and is easily propagated by cuttings: the best method of pruning is to leave but few of the lowest canes, which should be trimmed two or three feet long, and gathered together and tied to a stake three feet high. With proper care the fruit may be kept in a state of good preservation for months after gathering.

The superb clusters of the Emperor will probably average the very largest grown in the vineyards of California; in some instances attaining a length of from twenty-three to twenty-four inches, and a weight of from six to seven pounds. The small fruit-stems cling firmly to the berries, even when the fruit is perfectly ripe: this fact, added to the facility with which the bunches may be packed, their great size, remarkable beauty and agreeable taste, together with the great productiveness of the vines render the Emperor one of the very best of market table grapes.

The specimen represented was taken from a vineyard in Yolo County.